ROCK CITY BARNS

A Passing Era

ROCK CITY
BARNS
A Passing Era

Best Wishes!
Dave —

thanks for
your help!
Tom Snow
T. J. Snow Co.
4-29-97

Photographs and Text by David B. Jenkins

Free Spirit
PRESS

FIRST EDITION

Library of Congress Catalog Card Number: 96-096410

ISBN: 0-9652308-0-5

Published by:
Free Spirit Press
730 Cherry Street, Suite J
Chattanooga, Tennessee 37402

Printed in Belgium
10 9 8 7 6 5 4 3 2 1

BOOK DESIGN AND PRODUCTION
Westcott Design Group, Chattanooga, Tennessee

Design and project coordination by Kathryn E. Westcott
Color separations by Daniel Westcott
Special Thanks to Sissy Moore

DEDICATION

For Louise,
who perseveres...

Earth is crammed with heaven
And every bush aflame with God.
But only those who see
Take off their shoes.

—Elizabeth Barrett Browning

ACKNOWLEDGMENTS

A debt of gratitude is owed to each of the many individuals who played a part in making this book a reality. I am especially grateful to the following people:

To Bill Chapin, President of See Rock City, Inc., who dreamed of a book of Rock City barns—and then made the dream come true.

To Todd Smith, Marketing Manager for See Rock City, Inc., whose friendship and cooperation made the nitty gritty details of the project a pleasant experience.

To Kathryn Westcott of Westcott Design for a beautiful design and for constant encouragement, without which this book might never have existed.

To Fritz Henle, Elliot Erwitt, Robert Doisneau, and B.A. King: photographers who taught me to see beauty in the commonplace.

To Louise Jenkins, my wife, who blessed my long hours and frequent absences with unfailing sweetness of spirit.

This book is for all those who love the good things of America's past. But more than anything else, this book is for the barns themselves. ■

PREFACE

This book had a long gestation period. In 1988, Bill Chapin, president of See Rock City, Inc. and great-nephew of founder Garnet Carter, told me of his long-time dream to create a book about Rock City's barns and asked me to find out what it would cost. He decided not to proceed at that time, but my interest had been kindled. I obtained a list of the 110 barns they were still painting, and whenever my travels brought me near one I made a photograph of it if possible.

In 1994, after learning that the number of barns being repainted had dwindled to 85, I went to Chapin with some of my photos and told him I felt that if we were ever to do a book, now was the time. He didn't say much. Just looked at the pictures for about 15 minutes, asked a few questions, then said, "Let's do it!"

In a few days he sent me a box containing hundreds of old office file cards from the 1960s, Rock City's only record of most barn locations. On each card was the name of the property owner at that time, the highway, and the distance from the nearest town. Many had a small photo attached, apparently taken about 1960; but some had only rough sketches of the barns. Inside the fold-over card was a record of rents paid (usually $3 to $5 per year) and repaint dates. Rock City had had no contact with most of these barns since the late 60s. The only way to find out if they were still standing was to go and see.

Sorting the cards into piles by states (15), and within states by highways, I planned an itinerary and began photographing at Sweetwater, Tennessee on October 24, 1994. Over the next 16 months, stealing time from my studio whenever I could, the trail of barns led my old Chevy Blazer nearly 35,000 miles to more than 500 sites. Nearly 255 barns were found in 14 states: Alabama, Arkansas, Georgia, Illinois, Indiana, Kentucky, Louisiana, Mississippi, Missouri, North Carolina, Ohio, South Carolina, Tennessee, and Texas.

Each state is represented by at least one color photograph and every known barn is pictured, including nearly 20 of which Rock City had no record. Some of the forgotten barns were located through leads from local residents; others were discovered en route to known sites. Still more undoubtedly exist, lost in the musty recesses of Garnet Carter's desk-drawer filing system. It was always exciting to round a bend and see a Rock City barn for the first time—but immeasurably more thrilling to find one which had slipped through the cracks of history.

Not all Rock City barns, by the way, are barns. Sheds, garages, roadside stores, even a few abandoned houses and lone-standing silos have worn the familiar logo over the years. In fact, a third of all Rock City barns may have been something other than barns.

With 35-year-old, often sketchy records and occasional hearsay reports as my only sources of information, finding the sites was endlessly fascinating detective work. Barns have burned, blown down, been bulldozed for highway construction and subdivisions, or simply fallen from disuse and disrepair, sagging silently into the soil. Many of the largest and finest barns are gone. To complicate things still further, highways have been changed, re-routed, and re-named.

Often, the only way to locate a site was to find someone who remembered the property owner:

"Did you know so-and-so, who had a place out on Highway 11 south of here?"

"Oh, yeah, knew him well. He and my daddy used to go fishing together all the time. Good ol' feller. He's dead now."

"Well, he had this barn on his farm, with a sign that said 'See Rock City.' Here's an old picture of it."

"Sure, I remember that ol' barn. Fact is, I helped him take it down, back around 1975. It had got all rotten and falling down, y'know. Wasn't safe."

An average day of photography would involve driving over 400 miles in 12 to 15 hours, and would result in locating eight or nine sites, of which three to five might have barns. Some days were better than that, of course, and some were much worse. I began, I suppose, with some idealism, even though I'm a country boy and should know better. Expecting to find prosperous, story-book barn yards, I sometimes found depressing scenes of rural desolation.

What began as a commission became a labor of love as I grew to treasure the dignity and individuality of each old structure.

I learned to see beauty even in the isolation in which so many barns are ending their days. As a veteran photographer, I know most of the ways to help subjects look their best. But I soon learned that the barns wanted to be photographed in a simple, direct, even documentary way, without artifice. They seemed to say, *"Here we are. This is the way we are. Please let us speak for ourselves."*

So here they are in a book of their own. You will hear what they have to say if you listen with your eyes. And if looking through these pages gives you a fraction of the pleasure I've gained in preserving their memory, we both will be blessed.

—Dave Jenkins

THE MAN
WITH THE
PAINTBRUSH

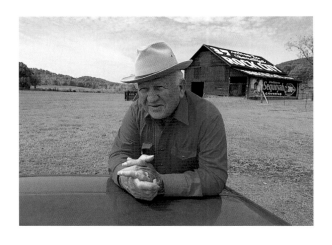

At eighty, Clark Byers looked remarkably fit and not at all like someone who had spent his life in advertising. But even though he never wore a gray flannel suit or came within hog-calling distance of Madison Avenue, Byers was the key player in one of the greatest advertising campaigns of its time; a venture which made an obscure tourist attraction near Chattanooga, Tennessee world famous and its slogan a household phrase. From the thirties through the sixties, Clark Byers traveled tens of thousands of miles in nineteen states and painted "See Rock City" on so many barns that no one knows to this day exactly how many there were.

"I was just a kid in my early twenties, working as a sign painter's helper. My boss was Mr. Fred Maxwell — he was real close friends with Garnet Carter, the man who started Rock City. Well, one day about 1936 or 37 Mr. Maxwell asked me to come up on the mountain with him to meet Mr. Carter and that was the startin' of it all."

Carter hired Clark to contact farmers and tell them that he would paint their barn roofs free if he could put a message on the roofs. "What's the message?" Byers asked.

Carter thrust a piece of paper across the desk. On it were scrawled three words: "See Rock City."

In the beginning, Carter and Maxwell chose locations for the signs. But in Byers' view, some of their selections were not good. "Sometimes I'd go up to the farmer's door sayin' to myself, "Nobody home, I hope, I hope, I hope!

"There was an ol' roof down at Tunnel Hill, Georgia, about 75 feet long. They hadn't ever seen it. They'd been by there, but it was settin' in behind some trees. I knew if I could cut the trees down it would knock their eyes out, so I got permission to cut down three or four trees and paint that barn. You could see it at least three quarters of a mile—that's how good it was. It was so much better than anything they had ever picked out, and they didn't even know it was there. So from then on, they just turned me loose and let me pick the sites."

Traveling around the southeast and up into the midwest in a pickup truck on trips of two or three weeks duration, Byers worked year 'round to get out the Rock City message.

"We tried to do the northern signs in the summertime and the southern signs in the wintertime, but it wouldn't always work out that way. Sometimes it would get so hot up there it would melt the crepe soles on our shoes.

"There's been a time when we would go out and do five or six a day. Three or four wasn't nothin'. I started out with one helper, then I got two helpers. Everybody had somethin' to do—we didn't miss a lick! These boys, I had 'em trained. We was hungry, and in a hurry."

Even working rapidly and with basically primitive methods, Byers' lettering was remarkably precise. In fact, although he's seen paintings of his barns by many artists, he says he has never seen one that got the lettering just right.

"I did it all freehand, y'know. That's the reason I could do 'em so fast and cover so much country. I never measured nothin' off, except the lines—we would stretch 'em across the roof with nails and pieces of string to mark it off. We'd black out the whole roof with black paint and then come back and put the white letters on it.

Sprayin' was a no-no! Not fast enough. We did it all with a four-inch brush. In the beginnin', we mixed our own paint with lamp black and linseed oil. Once that paint got on, there was no gettin' it off.

"Garnet Carter, he didn't argue about how many you got. It was how good you did 'em and who all saw 'em, y'know. He started out payin' me forty dollars a sign, and after that he got to payin' me by the foot. Ever' night I'd send Mr. Carter a postcard with sketches of the signs we'd done. He'd just stick things in his desk. He wasn't much on record-keepin'."

For Clark and his crew, life on the road was seldom dull. In Mississippi, an angry bull refused to allow them to come down from the roof they were painting. In Wisconsin, union men tried to keep Byers and his non union helpers from working. And of course, falling was a constant hazard. If a painter slipped, there was only one way to keep his balance and avoid a serious fall: "We'd be on the barn roof, y'know—I was pretty active back then, fifty years ago—but you could start slippin'. Once I slipped, I'd start runnin'(to stay upright) and I would jump letters. The paint on the letters would still be wet. The black background was dry, but if you stepped in that white, you was gone. So I'd run down through there jumpin' those letters (toward the lowest point on the roof), and naturally when I got to the edge I just had to jump to the ground to keep goin'. I'd hold that paint out there and balance it where it wouldn't spill all over ever'thing. If we spilled a bucket of paint in the man's lot, we'd have to dig it up. Kills the cattle, y'know?"

Even for a tee-totaler, there were other pitfalls: "There was this guy over at Kimball, Tennessee name of Earnest Smith. He owned the first barn I ever painted. One day we were repaintin' his barn, and he wanted to show me something, so I went down to his ol' smokehouse with him, and he had it full of wine. He'd made several different kinds of wine. Blackberry wine, cherry wine, you name it. He poured me out a cup, 'Hey, try this right here.' Well, I was tryin' to keep that sign. It was a good sign. You know how a guy can put pressure on you, and before he got done with me I had drunk at least a half a dozen cups of wine. I found out before I got the sign done that I wasn't gonna be able to drive. So my helper drove the truck home and I slept the whole way. That's the first time I ever got a little topsy (sic)."

In 1947 Clark built the mountain-stone house on a hundred acres of land on U.S. 11 near Rising Fawn, Georgia where he raised his family of three boys and two girls—and painted "See Rock City" on the roof.

Life on the road was rewarding, but sometimes a man could get homesick: "I got to where I enjoyed, y'know, takin' these trips and meetin' people out on the road. We had a lot of fun, me and my helpers. I would dread goin' out on the road so bad till I got about fifty miles away from home. Once I got about fifty miles, it was all over then. I'd work my head off to get back home. One time we were up on the Tennessee-Kentucky line and I wrote Mr. Carter a postcard:

'Out of paint and out of money,
Going home to see my honey."

By the 1960s, things were changing. The rapidly-growing interstate highway system drew long-distance travelers away from the old routes where most Rock City barns were located, and the Highway Beautification Act of 1965 (the "Ladybird law") forced Rock City to stop advertising on many of the barns. Although nearly 260 were still standing in 1996, only about 85 were still being maintained by the attraction. As for Byers, his career as a painter came to an abrupt end in 1968.

"I was paintin' a billboard at Murphreesboro (Tennessee) on U.S. 41. For some reason this power line had tore loose from the pole and was droopin' down over the sign. Well, I'm used to seein' stuff hangin' around like that—I didn't think too much about it, y'know . There was a big place on the sign where the paint had peeled and I was scrapin' it with a putty knife, and all of a sudden, BAM! Right in my ear—loud as a shotgun! It set my hair on fire and burned a place on my back—It was just a miracle it didn't kill me, y'know. A big truck had passed by and the blast of air from it swung that line up against me. They claimed it was seventy-two hundred volts! I was in the hospital a while and couldn't do anything for a year. It was just one of God's miracles I didn't die right there on the spot."

A warm, soft-spoken man of deep Christian faith, Byers is quietly happy with the life he has lived. In addition to his years of painting Rock City barns, he was the developer of Sequoyah Caverns, a tourist attraction in northeast Alabama, and at the age of eighty still ran a three hundred acre farm.

He also invented the red-and-black birdhouse which is now a principal advertising device for Rock City.

"I painted the first birdhouse that Rock City ever had. It was my original idea—I didn't copy nobody. I was gonna use it for a mailbox, but the Post Office wouldn't let us. It's still settin' in my garage.

"It's been kinda crazy, but I have enjoyed my life. I'm a strange person. I do things that other people wouldn't think about doin'. It's kinda always been that way. I don't copy nobody." ∎

SEE ROCK CITY

Yes, Virginia, there really is a Rock City. But the question is not irrelevant, because Rock City is a place which has long been overshadowed by its own advertising. Millions of people have seen Rock City, but so many millions upon millions more have read its slogan on barns, billboards, and birdhouses that "See Rock City" has taken on a life of its own and passed into American folk history. No wonder some people ask if there really is a Rock City, or if it's like that other great slogan from the mid-century, "Kilroy was here."

Probably the third-most-asked question by travelers south in the 1950s and 60s (after "Are we almost there yet?" and "When do we eat?") was "What the heck is a rock city?" Tempted beyond endurance by the pervasive message of the ubiquitous signs, millions made the pilgrimage up Lookout Mountain to see for themselves what it was all about.

What they found was a ten-acre tract of massive stone formations (hence the name "Rock City") on the eastern cliffs of the mountain overlooking Chattanooga, Tennessee and the North Georgia countryside. Laid out with flower-bordered

and along cliff tops, Rock City Gardens is an oddly pleasing blend of magnificent natural beauty and a charming naivete. The naivete comes in the form of trailside elfin figurines, walking "storybook" characters (Mother Goose, Humpty-Dumpty, Rocky the Rock City elf), and Fairyland Caverns, a man-made "cave" featuring black lighted dioramas of nursery rhyme scenes. This is what family entertainment was like before Disney World and video games, and the place still holds a special fascination for kids of all ages.

Although the rock formations have been there for a long, long time, Rock City as a tourist attraction is the brainchild of a freewheeling, cigar-smoking entrepreneur named Garnet Carter and his wife Frieda, a lover of beauty and folklore. During the 1920s, while Garnet was developing a fashionable 300-acre subdivision called Fairyland on Lookout Mountain and inventing the game of miniature golf (which he franchised nationally under the name "Tom Thumb Golf"), Frieda was making Rock City, which was a part of Fairyland, into her own personal nature garden. She laid out trails and planted their borders with hundreds of wildflowers and shrubs indigenous to the southern mountains. For her efforts she was awarded the Bronze Medal of Distinction by the Garden Club of America in 1933, the first Southerner to be so honored.

In the meantime, the Carter fortunes had changed. The stock market crash of 1929 put an end to the building boom of the 20s. And while Tom Thumb Golf was doing very well, Carter foresaw the end of the miniature golf craze and managed to unload in 1930 at a handsome figure. Unfortunately, he invested the proceeds in U.S. Steel stock at $298 a share, which promptly dropped to $30.

Finding himself not exactly broke, but feeling some pain, Carter turned his full attention to his wife's little garden. Motivated both by economic possibilities and by love for his wife, who was then in the beginning stages of a debilitating disease, he began to expand upon the improvements she had begun. The trails were widened and paved with flagstones and suspension bridges were built over chasms. Statues of gnomes and storybook characters, which had adorned the Fairyland neighborhood and the original Tom Thumb golf course at the

Fairyland Inn (now the Fairyland Club) were moved to Rock City. On May 21, 1932 the park was opened to the public.

Unfortunately, the country was in the depths of depression and the tourism industry was still in its infancy. In spite of intense promotion, only eighteen hundred people came that first year, and things did not improve substantially until 1935, 36, or 37, depending on whose memory you trust. Clark Byers says 1937—or maybe 1936—was the year he began painting "See Rock City" on barns around the Southeast, and things began to look up for the attraction.

"I couldn't figure out how in the world Garnet Carter could afford to pay me $40 a sign," says Byers, "But boy, I went up to Rock City one day and the parking lots was full and people parked all around, and the signs had brought them in there. And I realized really, how smart he was, y'know."

Whether one can really see seven states from the sheer cliffs of Lover's Leap will continue to be be debated, but

Rock City is a place that's easy to like. 'Most all the half million folks who come each year feel the tour is well worth the price of admission. In fact, more than half of them are visiting for the second, third, or fourth time. You've seen the advertising. Now come See Rock City! ∎

THE BARNS

Kentucky 91 in Caldwell County, Kentucky

Alabama 202 (old U.S. 78) in Calhoun County, Alabama

In the early 20s, when Clarence Spindler was born on his father's farm in Gibson County, Indiana, U.S. 41 was a dirt track snaking through the countryside. The barn, built before Clarence was born, became a Rock City landmark in 1949, just two years after he brought his new wife home to the old farmhouse.

The wife has passed on now, and Clarence gave up farming a few years ago. "Not much fun anymore." Someone else tends the old place these days, and Highway 41 flows towards Florida on four smooth lanes of asphalt. The "See Rock City" sign on the barn was last repainted in 1967.

Like Clarence, the barn and its sign are showing just a little bit of wear. But they're hangin' in there pretty good, considerin'.

U.S. 41 in Gibson County, Indiana

M artha Shaddix always wanted to own a Rock City barn. Growing up in Florida, she rode up through the South in the back seat of her parent's car to visit relatives in Georgia and North Carolina. Watching the road, watching the barns go by, dreaming of a barn of her own. With a See Rock City sign, by golly.

Her dream came true in 1978 when she bought the old store at Halfway, Kentucky, complete with Rock City sign. Opened in 1923, the Halfway Trading Post is one of only two country stores still in operation bearing the Rock City message. The other is in West Tennessee and is not currently maintained by Rock City.

Life is kinda slow in Halfway, except on U.S. Highway 231, where traffic is kinda fast. Martha opens up her store every morning, six days a week, and since she's the Postmaster too, she goes out and raises the flag first thing.

U.S. 231 in Allen County, Kentucky

U.S. 127 in Bledsoe County, Tennessee

U.S. 41-A in Webster County, Kentucky

Photography is about light. In fact, the word "photo-graph" comes from two Greek words: photos, which means "light," and grapho, which means "to write." So to photograph really means to write with light.

The best light for photography usually comes in early morning and late afternoon when the sun is low, casting long shadows which reveal texture and bathe the land in a rich, golden glow. In summer this often meant 18-hour days, as I tried to capture the good light at both ends of the day. I might drive several hundred miles checking out barn sites and then double back to photograph an especially good one in evening light; or put up at a nearby motel if I thought morning light would be better.

Of course, there wasn't always a motel nearby. I drove 40 miles through the pre-dawn darkness of central Georgia to catch a June sunrise at this little barn which had been lost from Rock City's records. I learned about it from the neighbor who works on my car.

U.S. 19 in Taylor County, Georgia

U.S. 31 West in Simpson County, Kentucky

U.S. 129 in Graham County, North Carolina

U.S. 64 in Wayne County, Tennessee

Interstate 24 in Marion County, Tennessee

U.S. 11 in DeKalb County, Alabama

U.S. 11 in Clarke County, Mississippi

My longest barn photo expedition was an eight-day slog under gray and dripping March skies to 30 sites scattered around Arkansas, Louisiana, Mississippi, and Texas. I found only six barns. One of the most attractive is this small "mule barn" in Northeast Louisiana, which was first photographed under a leaden sky.

The barn deserved better, but time was tight. So I pushed ahead on a two-day zig-zag across Mississippi which brought me empty-handed to nightfall at Greenville, 60 miles from the Louisiana barn. With the forecast still forbidding, I resolved that sunrise, if it came, would find me back at that barn.

I was a little behind schedule, but not the sun. It beat me to the spot by ten minutes and radiantly blessed my efforts, including the close-up of sun-streaked blackberry leaves and weathered boards which grace the end papers of this book. *And then I heard the rumble of a distant train...*

Good things happen when you put yourself in position for them to happen.

U.S. 165 in Morehouse Parish, Louisiana

THE LAST LITTLE ROCK CITY BARN IN TEXAS

Most folks are surprised to learn there were once Rock City barns in Texas. Rock City had three on their records, but Clark Byers says there were more, including one near Tyler that belonged to an oil millionaire who was so cheap he wouldn't have indoor plumbing. But that was a long time ago, and Clark doesn't remember the location of any of them.

So, as far as can be known at this time, the last Rock City barn in Texas is on Richard Haynes' little spread on U.S. 80 west of Marshall. Richard and his wife Vallie (who was born on the place) are getting along in years now, but he still runs 17 head of registered Black Angus cattle on his 67 acres and keeps the big yard mowed—practically a full-time job in itself.

This is his second barn to carry the "See Rock City" message. The first was one of traditional design which was pulled down in the early 60s and replaced with the present "pole barn." Rock City obligingly applied their sign to the new barn as well, and that's the way it's stayed. It's pretty faded now, as it well should be—it was last repainted around 1967.

U.S. 80 in Harrison County, Texas

A barn is the perfect example of the architectural dictum that form should follow function. Need to store a lot of hay? Simple, just build a large loft with a gambrel roof. If you need more room, add on a wing or two. Barns for horses or mules, on the other hand, usually have small lofts, or none at all, and are divided into stalls, runways, and storage rooms for feed. Tobacco curing barns vary according to region, but in Kentucky, which seems to have more of them than anyplace else, they tend to be longish and narrow with gable roofs and hinged sideboards that swing out like shutters to let the air flow through freely. (A fine example is on page 55.)

Southern barns tend to be smaller than Northern barns. Shorter winters mean less hay-storage capacity is needed, and livestock can usually stay outside. In mountain areas the land is often not very productive and the people poorer, so they seldom build large barns. But in prosperous farm country like Pennsylvania and Ohio you can find magnificent barns more than a century old.

U.S. 11 in DeKalb County, Alabama

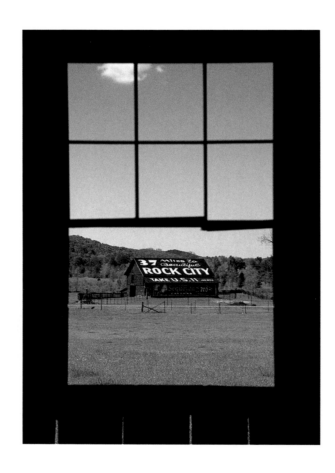

U.S. 11 at Sequoyah Caverns in DeKalb County, Alabama

Alabama 43 in Washington County, Alabama

U.S. 27 in Scott County, Tennessee

The Federal Highway Beautification Act forced Rock City to stop painting many barns. Worse, they were required to obliterate many signs by painting over them.

Fortunately the paint-overs didn't weather well. On this barn the red has worn away from the letters, making it the only red-and-white Rock City sign around. Last I heard, they planned to touch it up with some fresh, new white.

The Highway Beautification Act has made many stretches of highway, especially interstates, terminally boring. I sometimes see people reading books or magazines as they drive. They would be safer reading signs. And as for beauty, what could be more beautiful than a Rock City barn?

Besides, I miss the Burma-Shave signs.

U.S. 19 in Swain County, North Carolina

U.S. 41-A in Rutherford County, Tennessee

U.S. 31 East in Barren County, Kentucky

U.S. 11 East in Hamblen County, Tennessee

U.S. 11 in Louden County, Tennessee

Old U.S. 431 in Randolph County, Alabama

U.S. 70 North in Smith County, Tennessee

U.S. 31 East in Hart County, Kentucky

U.S. 129 in Irwin County, Georgia

U.S. 41 in Christian County, Kentucky

Tennessee 68 in Cumberland County, Tennessee

Interstate 40 in Roane County, Tennessee

U.S. 441 in Sevier County, Tennessee

U.S. 129 Blount County, Tennessee

Something draws us to old barns. Painters paint them, photographers photograph them, writers write about them. Tourists slow down when they pass an especially fine one so they can take a good look at it. On the plains of central Ohio, farmers of another generation put their names on the great barns they built so the world would know of their achievements.

A barn is a simple answer to simple needs: a place of shelter for domestic animals, and a storage place for food. Perhaps we think they are beautiful because of their simplicity and utility, or perhaps it's because they remind us of a less complicated age when life moved at the pace of the seasons rather than at the pace of a time-clock.

A friend of mine says old barns have souls. Could be he's right.

U.S. 64 in Davie County, North Carolina

Strange as it may seem, not all of Rock City's barns have been faithful to their painter. This barn near Manchester, Tennessee is one of two turncoats which have transferred their allegiance to Ruby Falls, another Lookout Mountain tourist attraction. They at least have not yet lowered themselves to the point of advertising chewing tobacco.

To give credit where it's due, it was the chewing tobacco folks who started the whole thing. Bloch Brothers Tobacco Company of Wheeling, West Virginia began painting their Mail Pouch Chewing Tobacco signs on barns in 1897. Some others who got into barn advertising in a big way are Jefferson Island Salt, Ruby Falls (mostly in Tennessee), and Merrimac Caverns of Missouri, which has barns all over the Midwest.

U.S. 41 in Coffee County, Tennessee

U.S. 41 in Grundy County, Tennessee

U.S. 11 West in Grainger County, Tennessee

U.S. 31 East in Spencer County, Kentucky

U.S. 31 in Lowndes County, Alabama

U.S. 150 in Washington County, Indiana

U.S. 64 in Cherokee County, North Carolina

U.S. 411 in Bartow County, Georgia

U.S. 129 in Blount County, Tennessee

U.S. 11 in DeKalb County, Tennessee

Old U.S. 62 in Brown County, Ohio

U.S. 231 in Wilson County, Tennessee

Old U.S. 68 in Nicholas County, Kentucky

You may not have noticed it, but barns are actually something of an endangered species these days. They don't build 'em like they used to, and one reason is the way farmers feed their livestock. The traditional barn was built with a large loft to store loose hay. The farmer went up there with a pitchfork and tossed hay down to his animals. Then square bales were invented, and that was okay, because they would also fit in the barn loft, and since they were tightly packed you could store even more hay. But if something happened to the barn there was not much reason to replace it because hay can be stored more conveniently in an open, metal-roofed shed built at a fraction of the cost of a barn.

And then, along came the big, round bales. No way you're going to get those 1000-pound babies in a loft. They have to be handled by a tractor with a front-loader. So nobody much builds regular barns anymore, unless they have dairy cattle or horses. And the good old barns are falling down and not being replaced.

U.S. 441 and 129 in Morgan County, Georgia

U.S. 41 in Gordon County, Georgia

U.S. 31 West in Barren County, Kentucky

OUTLASTING

THE

PAINT

like the faded barns best.

That lampblack and linseed oil paint Clark Byers and his crew slapped around was tough stuff. I've seen barns which appeared to be held up by the Rock City paint and not much else. It lasted better on boards than on tin, and best of all on the north-facing end of a barn. Clarence Schindler's barn hasn't been redone since 1967, but the paint is still in remarkably good shape.

Metal roofs eventually rust, even with a good coating of Clark's paint, but the black background and white letters oxidize in different ways so a sign is often quite readable in two different shades of rust.

Most interesting are the barns which have been repainted over the years with different messages that have faded into each other. This is one: can you read it? Another is on page 39.

U.S. 41 in Todd County, Kentucky

U.S. 25 in Harrison County, Kentucky

U.S 11 West in Grainger County, Tennessee

U.S. 27 in Miller County, Georgia

Old U.S. 411 in Murray County, Georgia

When interviewing master barn painter Clark Byers, his wife was present. I asked her, "Mrs. Byers, was your husband a drinking man?"

"Oh, no! Never! He never drinks."

"But what about the days when he was on the road for weeks at a time? Are you sure he didn't sometimes kill a few beers with the boys at the end of a hard day?"

"No — no, he just wouldn't do that."

"Then — (showing her a picture of the scrambled barn) how do you account for this?"

In all fairness to Mr. Byers' reputation for sobriety, let it be noted that he didn't paint the barn this way. The owners removed one wing, then used the roofing metal to overlay the exposed side without regard to Clark's carefully laid-out sign.

U.S. 82 in LaFayette County, Arkansas

It often took longer to find a barn than it did to photograph it. It might be gone, of course. More than half of them are. And if it still stands it might be weathered and faded to indecipherablilty, the sign may have been painted out — a few have even been overlaid with sheet metal — or a wing may have been added or removed.

This one would have been just about impossible to locate without the little photo on the old Rock City file card. The photo showed, in addition to a small barn which looked nothing like this one, a large ante-bellum house which the folks at a store down the road recognized immediately as the "1850s House."

One wing of the barn has been removed and most of the end swathed in sheet metal. If you look carefully, you can see the right side of the "O" and the back of the "C" in "Rock."

U.S. 41 in Lamar County, Georgia

U.S. 441 in Sevier County, Tennessee

U.S.11 in Monroe County, Tennessee

U.S. 40 in Warren County, Missouri

South Carolina 28 in Abbeville County, South Carolina

U.S. 231 in Lincoln County, Tennessee

Old U.S. 70 in White County, Tennessee

LOST

IN THE

WILDWOOD

A lot can happen in 35 or 40 years. Trees can grow up, barns can fall down, highways can be moved. Whole populations of people can move into or out of an area. Central and southeast Georgia are like that; the roads are lined with abandoned houses, many of them surrounded with trees and brush. You can see lots of people used to live there, but where have they all gone? The barn on page 109 is a few miles from Dublin, in a region that God probably hasn't forgotten about, but a lot of other folks appear to have.

The Georgia barn on the facing page took several hours to find. Searching along U.S. 341 near Brunswick, I passed it numerous times before finding someone who knew exactly were it was. Arriving at the spot, nothing could be seen except a dense thicket. I parted the foliage with my hands and there it stood, just as you see it here.

U.S. 341 in Glynn County, Georgia

Old U.S. 11 East in Greene County, Tennessee

Georgia 42 in Monroe County, Georgia

U.S. 150 in Orange County, Indiana

U.S. 80 in Laurens County, Georgia

ENDANGERED SPECIES

From 1950 to 1990, the number of farms in this country dropped from more than five million to only two million, while we went from 23 million farm families to fewer than six million. Yet at the same time, farm production doubled and the size of the average farm rose from 215 to 450 acres. A farm has to be a pretty good sized operation to support a family. The small family farm—and its barn—are becoming anachronisms.

Agri-businesses don't build many barns. For what they do, metal utility buildings are more practical. If they're also ugly, that's beside the point. Business is business.

So enjoy these barns. There aren't many more where they came from.

U.S. 60 in Ballard County, Kentucky

The photograph at left was made on a memorable May day in 1994. The one at right is the same barn in December, 1995. If you look carefully at the upper center of the picture, you can see the reason for the destruction — approaching highway construction. The four-laning of U.S. 27 between Rockwood and Spring City, Tennessee caused the loss of two Rock City barns that I know of, and I suspect there may have been others.

The little cupola which once crowned the ridgepole sits intact on the ground at center right.

No one keeps statistics on this sort of thing, but probably as many barns are lost to highway construction and expansion as to any other hazard. But, hey, who cares? It's just an old barn, right?

It would be nice if some of these barns could be preserved as historic landmarks.

"To miss Rock City would be a pity..."

U.S. 27 in Rhea County, Tennessee

U.S. 31 East in Hart County, Kentucky

U.S. 50 in Clay County, Illinois

U.S. 41 in Coffee County, Tennessee

U.S. 64 in Giles County, Tennessee

Clark Byers invented the Rock City birdhouse. He thought it would make a nice mailbox, but the Post Office didn't agree. Somewhere along the line the Rock City folks picked up on the idea, and now they sell thousands of them every year.

In 1967, Clark and his oldest son Freddie painted the barn on the old Tibbs farm in Dalton, Georgia to look like a giant birdhouse. It's less than a hundred yards from Interstate 75, and would be a spectacular sight if you could see it. Unfortunately, the government types who regulate such things won't allow a few trees to be cut, so you'll have to take my word for it. Or if you stop one-tenth mile south of milepost 335 on the northbound side and look to your right, you can see it through the foliage.

I have nothing against trees, but I like to look at Rock City barns too. Don't you? Write your congressman.

Interstate 75 in Whitfield County, Georgia

U.S. 31 West in Hart County, Kentucky

U.S. 11 in DeKalb County, Alabama

U.S. 231 in Madison County, Alabama

Interstate 24 in Marion County, Tennessee

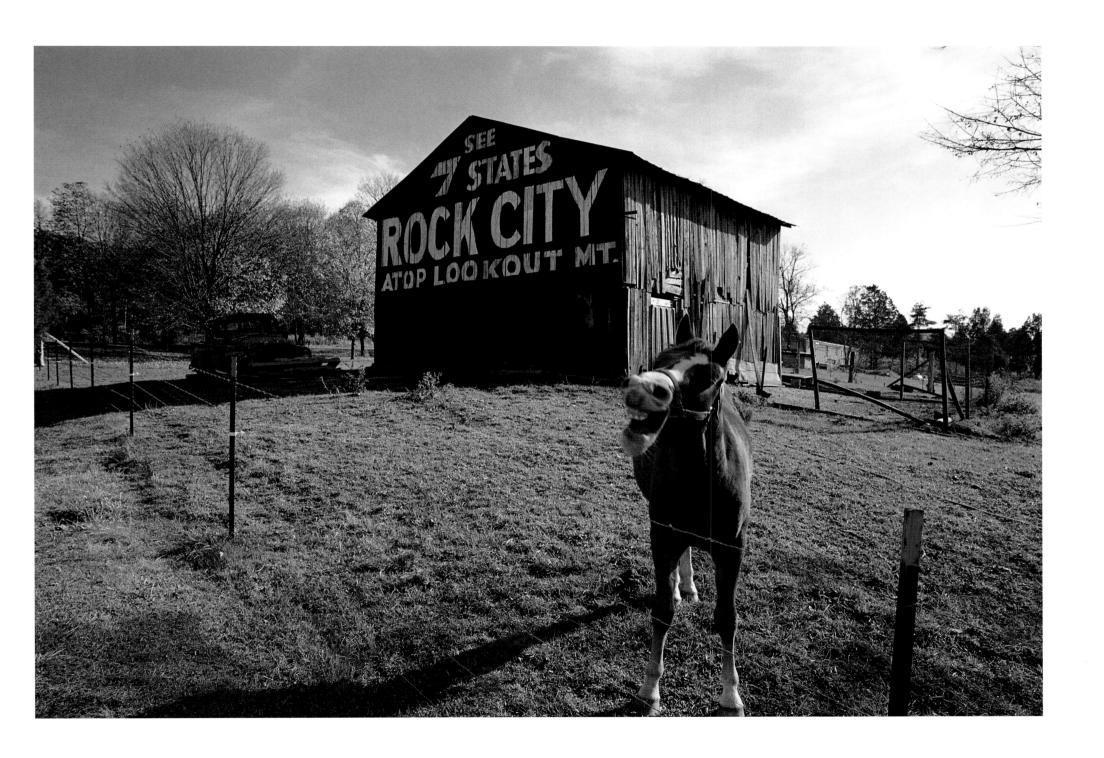

U.S. 11 in McMinn County, Tennessee

REMNANTS OF A PASSING ERA

I wish every barn could have a color page of its own. That isn't possible, of course. But at least they're all here: every Rock City barn we could find. Well-kept or abandoned, they have a charm about them that, lingering, links us with a vanishing way of life.

ALABAMA

Alabama 21 in Talladega County, AL

U.S. 431 in Chambers County, AL

U.S. 11 in DeKalb County, AL

U.S. 11 in DeKalb County, AL

U.S. 31 in Conecuh County, AL

U.S. 11 in St. Clair County, AL

U.S. 72 in Lauderdale County, AL

U.S 431 in Marshall County, AL

U.S. 11 in DeKalb County, AL

U.S. 11 in DeKalb County, AL

Alabama 43 in Greene County, AL

U.S. 11 in Greene County, AL

Alabama 5 in Bibb County, AL

U.S. 11 in Tuscaloosa County, AL

ARKANSAS

U.S. 70 in St. Francis County, AR

GEORGIA

U.S. 27 in Walker County, GA

U.S. 411 in Bartow County, GA

U.S. 27 in Troup County, GA

U.S. 80 in Twiggs County, GA

U.S. 80 in Twiggs County, GA

U.S. 441 in Wilkinson County, GA

U.S. 341 in Dodge County, GA

U.S. 341 in Pulaski County, GA

Georgia 42 in Monroe County, GA

Old U.S. 341 in Monroe County, GA

Georgia 48 in Chatooga County, GA

U.S. 11 in Dade County, GA

U.S. 11 in Dade County, GA

U.S. 11 in Dade County, GA

U.S. 11 in Dade County, GA (Clark Byers' house)

I-59 in Dade County, GA

U.S. 441 in Oconee County, GA

U.S. 80 in Emanuel County, GA

U.S. 341 in Crawford County, GA

U.S. 341 in Telfair County, GA

U.S. 41 in Tift County, GA

U.S. 23 in Bibb County, GA

U.S. 41 in Whitfield County, GA

U.S. 27 in Walker County, GA

ILLINOIS

U.S. 67 in McDonough County, IL

Illinois 125 in Cass County, IL

U.S. 67 in Greene County, IL

U.S. 67 in Fayette County, IL

U.S. 51 in Shelby County, IL

U.S. 51 in Fayette County, IL

U.S. 45 in Johnson County, IL

Illinois 142 in Saline County, IL

Illinois 250 in Richland County, IL

INDIANA

U.S. 231 in Owen County, IN

U.S. 31 in Jackson County, IN

I-65 in Clark County, IN

KENTUCKY

U.S. 41N in Christian County, KY

Kentucky Highway 1247 in Pulaski County, KY

U.S. 27 in Garrard County, KY

U.S. 27 in Garrard County, KY

U.S. 27 in Garrard County, KY

U.S. 27 in Harrison County, KY

U.S. 27 in Harrison County, KY

U.S. 62 in Robertson County, KY

U.S. 62 in Harrison County, KY

U.S. 25 in Rockcastle County, KY

U.S. 25W in Whitley County, KY

Old U.S. 31E in Allen County, KY

Kentucky 91 in Christian County, KY

U.S. 641 in Crittenden County, KY

U.S. 60 in Bedford COunty, KY

Kentucky 109 in Webster County, KY

U.S. 41 in Christian County, KY

U.S. 41-A in Christian COunty, KY

U.S. 31W in Simpson County, KY

U.S. 31W in Warren County, KY

U.S. 31W in Barren County, KY

Old U.S. 31W in Hardin County, KY

U.S. 31W in Hardin County, KY

U.S. 31E in Hart County, KY

U.S. 31E in Hart County

U.S. 31W in Hart County, KY

U.S. 31W in Barren County, KY

U.S. 60 in Carter County, KY

U.S. 68 in Nicholas County, KY

LOUISIANA

U.S. 90 in Calcasieu Parish, LA

MISSOURI

U.S. 40/I-70 in Warren County, MO

U.S. 40/I-70 in Montgomery County, MO

NORTH
CAROLINA

U.S. 17 in Beaufort County, NC

U.S. 17 in Jones County, NC

Old U.S. 1 in Wake County, NC

U.S. 64 in Randolph County, NC

U.S. 64 in Davie County, NC

U.S. 231 in Catawba County, NC

U.S. 74 in Cleveland County, NC

Old U.S. 64 in Clay County, NC

U.S. 19 and 129 in Cherokee County, NC

OHIO

U.S. 52/62/68 in Brown County, OH

U.S. 68 in Brown County, OH

U.S. 68 in Brown County, OH

Old U.S. 25 in Shelby County, OH

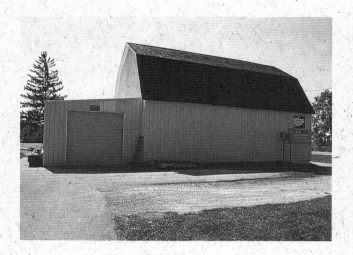

U.S. 127 in Mercer County, OH

Old U.S. 27 in Hamilton County, OH

U.S. 42 in Warren County, OH

TENNESSEE

Highway 68 in Cumberland County, TN

U.S. 11 in Monroe County, TN

Tennessee Highway 58 in Roane County, TN

I-40 in Roane County, TN

U.S. 70 in Roane County, TN

U.S. 27 in Morgan County, TN

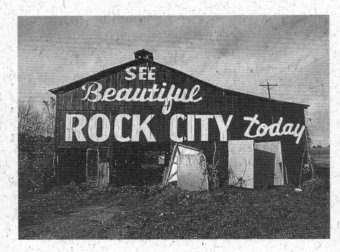

U.S. 11 E in Jefferson County, TN

U.S. 11E in Grainger County, TN

Old U.S. 11W in Hawkins County, TN

U.S. 11E in Greene County, TN

U.S. 11E in Washington County, TN

U.S. 321 in Cocke County, TN

U.S. 321 in Cocke County, TN

U.S. 11E in Hamblen County, TN

U.S. 11E in Hamblen County, TN

U.S. 11E in Jefferson County, TN

U.S. 411 in Sevier County, TN

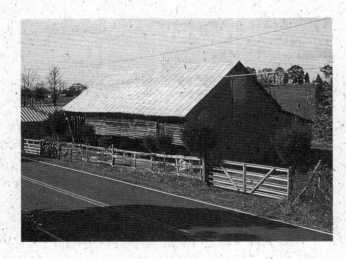

Old U.S. 411 in Sevier County, TN

U.S. 441 in Sevier County, TN

U.S. 441 in Sevier County, TN

U.S. 411 in Blount County, TN

U.S. 321 in Blount County, TN

U.S. 411 in Monroe County, TN

U.S. 231 in Rutherford County, TN

U.S. 231 in Wilson County, TN

U.S. 41-A in Bedford County, TN

U.S. 41-A in Rutherford County, TN

U.S. 41-A in Rutherford County, TN

U.S. 41-A in Bedford County, TN

U.S. 70 in Humphreys County, TN

U.S. 412 in Henderson County, TN

U.S. 70 in Haywood County, TN

U. S. 64 in Fayette County, TN

U.S. 64 in Wayne County, TN

U.S. 231 in Lincoln County, TN

Old U.S. 64 in Lincoln County, TN

U.S. 64 in Lawrence County, TN

Tennessee 20 in Lewis County

U.S. 231 in Lincoln County, TN

U.S. 64 in Lincoln County, TN

Old U.S. 70S in Cannon County, TN

U.S. 70S in Cannon County, TN

U.S. 70 in White County, TN

Old U.S. 70 in White County, TN

OLd U.S. 70S in Cannon County, TN

Old U.S. 70S in Warren County, TN

Tennessee 30 in Van Buren County, TN

Tennessee 30 in Van Buren County, TN

Tennessee 30 in Van Buren County, TN

U.S. 41-A in Davidson County, TN

U.S. 41-A in Cheatham County, TN

U.S. 41-A in Cheatham County, TN

U.S. 41 in Robertson County, TN

U.S. 41 in Robertson County, TN

U.S. 79 in Montgomery County, TN

U.S. 31E in Sumner County

Old U.S. 11W in Knox County, TN

I-24 in Marion County, TN

U.S. 41 in Marion County, TN

Tennessee 58 in Hamilton County, TN

Tennessee 283 in Sequatchie County, TN

Tennessee 27 in Marion County, TN

TECHNICAL NOTES

Most photographs for this book were made with Canon EOS cameras and Fujichrome film. Lenses (all EOS) were the 24mm f2.8, 28mm f2.8, 35mm f2, 50mm f1.8, 28-105mm f3.5-4.5 zoom. Most used was the 28-105. The front cover photo was made with a Mamiya RB67 and 90mm Sekor lens, again on Fujichrome film. Cokin filters in the "P" series were used as circumstances warranted; the 85A was used most frequently.

All pictures in this book are available as original photographic prints. Selected photos are also available as posters or as limited edition lithographs. For a free brochure and information about ordering photographs, call or write:

Free Spirit Press

730 Cherry Street, Suite J

Chattanooga, Tennessee 37402

423-265-4908 phone

423-265-4912 fax

Great care has been taken to include every barn which can be confirmed as having had a Rock City sign at any time. However, since the records are incomplete, we feel sure that some barns may have been missed. If you know of a Rock City barn that isn't in this book, please call or write the publisher so it can be included in future editions. If possible, send a color snapshot, along with the location and your name and telephone number. ∎